NATURE'S MYSTERIES

HURRICANES

KRISTI LEW

Britannica
Educational Publishing

IN ASSOCIATION WITH

ROSEN
EDUCATIONAL SERVICES

Published in 2019 by Britannica Educational Publishing (a trademark of Encyclopædia Britannica, Inc.) in association with
The Rosen Publishing Group, Inc.
29 East 21st Street, New York, NY 10010

Copyright © 2019 The Rosen Publishing Group, Inc. and Encyclopædia Britannica, Inc. Encyclopædia Britannica, Britannica, and the Thistle logo are registered trademarks of Encyclopædia Britannica, Inc. All rights reserved.

Distributed exclusively by Rosen Publishing.
To see additional Britannica Educational Publishing titles, go to rosenpublishing.com.

First Edition

Britannica Educational Publishing
J.E. Luebering: Executive Director, Core Editorial
Mary Rose McCudden: Editor, Britannica Student Encyclopedia

Rosen Publishing
Kathy Kuhtz Campbell: Senior Editor
Michael Moy: Series Designer
Tahara Anderson: Book Layout
Cindy Reiman: Photography Manager
Nicole DiMella: Photo Researcher

Library of Congress Cataloging-in-Publication Data

Names: Lew, Kristi, author.
Title: Hurricanes / Kristi Lew.
Description: First edition. | New York: Britannica Educational Publishing, in Association with Rosen Educational Services, 2019. | Series: Nature's mysteries | Audience: Grades 1–5. | Includes bibliographical references and index.
Identifiers: LCCN 2018008325| ISBN 9781508106623 (library bound) | ISBN 9781508106500 (pbk.) | ISBN 9781508106562 (6 pack)
Subjects: LCSH: Hurricanes—Juvenile literature. | Severe storms—Juvenile literature.
Classification: LCC QC944.2 .L49 2019 | DDC 551.55/2—dc23
LC record available at https://lccn.loc.gov/2018008325

Manufactured in the United States of America

Photo credits: Cover, p. 1 InterNetwork Media/DigitalVision/Getty Images; p. 4 Hector Retamal/AFP/Getty Images; pp. 5, 8, 9 NOAA; pp. 6, 15, 22, 24 Encyclopædia Britannica, Inc.; p. 7 Hindustan Times/Getty Images; p. 10 © AP Images; p. 11 Steven D. Starr/Corbis Historical/Getty Images; p. 12 Nicholas Kamm/AFP/Getty Images; p. 13 Win McNamee/Getty Images; p. 14 Nicolle R. Fuller/Science Source; p. 16 Ricardo Arduengo/AFP/Getty Images; p. 17 udaix/Shutterstock.com; p. 18 Paul Morse/The White House; p. 19 Jim Watson/AFP/Getty Images; pp. 20, 21 Designua/Shutterstock.com; p. 23 Inga Locmele/Shutterstock.com; p. 25 NASA; p. 26 Vitoriano Junior/Shutterstock.com; p. 27 vladimir salman/Shutterstock.com; p. 28 FloridaStock/Shutterstock.com; p. 29 Joe Raedle/Getty Images; interior pages background lavizzara/Shutterstock.com.

CONTENTS

WHAT IS A HURRICANE? ... 4
HURRICANE SEASON ... 6
STORM SAFETY ... 8
HURRICANE NAMES ... 10
BIG STORMS ... 12
HOW DO HURRICANES FORM? ... 14
WIND ... 16
WATER ... 18
OCEAN CURRENTS ... 20
CLIMATE ... 22
CLIMATE CHANGE ... 24
FUTURE HURRICANE SEASONS ... 28
GLOSSARY ... 30
FOR MORE INFORMATION ... 31
INDEX ... 32

WHAT IS A HURRICANE?

A hurricane is a type of storm. These storms are named for Huracan, a god of evil according to the people who once lived on islands in the Caribbean. The storms can be powerful and dangerous.

Hurricanes form over warm ocean waters. Their winds **spiral** around a central area. They bring heavy rains and strong winds

Hurricanes are powerful storms. They can cause a lot of damage, as Hurricane Maria did in Puerto Rico in September 2017.

VOCABULARY

To spiral means to circle around a center.

when they hit land. The rains can cause floods. The winds can destroy buildings and rip trees out of the ground.

Hurricanes have different names around the world. Any storm that forms in warm ocean waters and has winds that blow in a spiral is called a tropical cyclone. Tropical cyclones that form near North America and the Caribbean are called hurricanes. A tropical cyclone that forms near the Philippines, Japan, or China is called a typhoon. One that forms near Australia or in the Indian Ocean is often called a cyclone.

A satellite image taken in space shows the winds and rain spiraling around a central area.

HURRICANE SEASON

Most tropical storms form in the summer or early fall. The Atlantic hurricane season begins in June and ends in

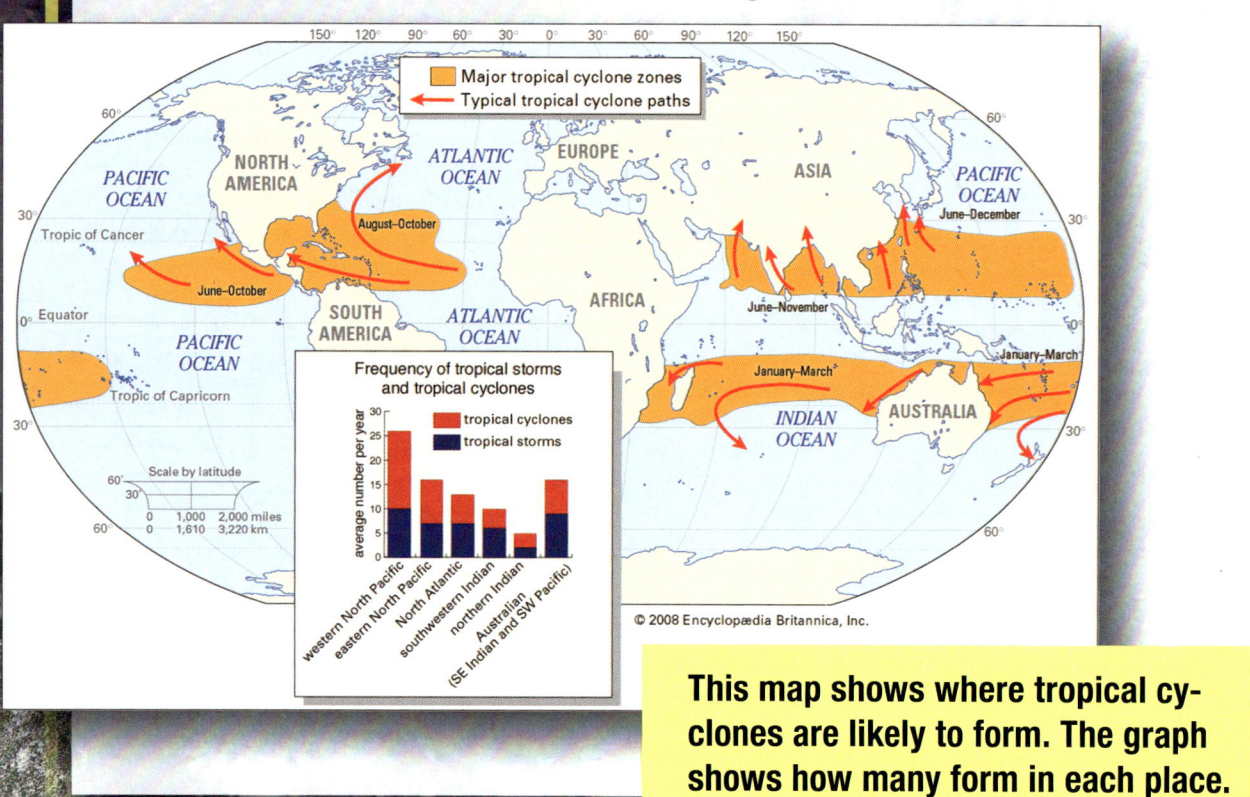

This map shows where tropical cyclones are likely to form. The graph shows how many form in each place.

THINK ABOUT IT

Why is it useful to know when hurricanes are most likely to happen? How could that information help you make plans?

November. Hurricanes are most likely to occur in September. In the Pacific Ocean, typhoon season is from May to October. The storms referred to as cyclones form in the Indian Ocean. The Indian Ocean lies mainly south of the equator, in the Southern Hemisphere. The seasons in the Southern Hemisphere are the opposite of the seasons in the Northern Hemisphere. Most cyclones form between December and April.

Hurricane season is the time of year when most hurricanes form. This does not mean that hurricanes cannot occur out of season. Some hurricanes form before June. Others occur after November.

Cyclone Ockhi brought wind, rain, and large waves to southern India on December 1, 2017.

STORM SAFETY

Scientists use satellites and airplanes to track tropical storms. They watch closely to see if the storms grow into hurricanes. The scientists warn people that a hurricane may hit a certain area. People in that area may need to board up their windows and seek shelter from the storm. They may even need to **evacuate**.

Scientists use their knowledge of weather patterns to judge where a storm might go next. They try to predict where the storm will

Satellite pictures allow scientists to see hurricanes forming. The pictures can help scientists judge where the storm might go next.

VOCABULARY

To **evacuate** means to leave a place of danger and go somewhere safe.

be up to five days into the future. The storm's track is most certain when it is close to an area. The forecast becomes less certain the farther away the storm is. The scientists use maps to show people where a hurricane might be headed.

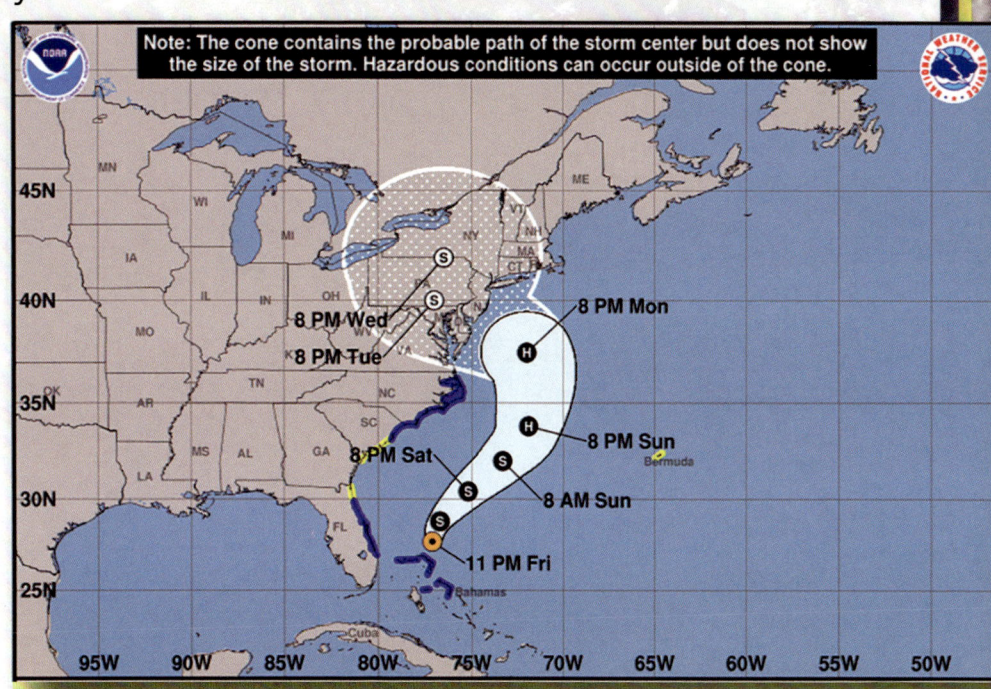

Note: The cone contains the probable path of the storm center but does not show the size of the storm. Hazardous conditions can occur outside of the cone.

Maps can show where scientists think a storm will most likely go in the future.

HURRICANE NAMES

Tropical storms and hurricanes are given people's names. An international group of scientists makes a list of the names to be used each year. There are separate lists for the hurricanes in the North Atlantic and those in the eastern North Pacific. The first tropical storm of the season in each list begins with the letter A, the second begins with B, and so on.

Sometimes two or more tropical storms or hurricanes can form in the Atlantic Ocean at one time. Short storm names that

A satellite image shows Tropical Storm Bonnie and Hurricane Charley on August 11, 2004.

VOCABULARY

Retired means that something has been taken out of use.

are easy to remember help to avoid mix-ups. The names help officials communicate with the public quickly and easily.

The United States began using female names for storms in 1953. Male names were added to the list for Atlantic hurricanes in 1979. Today, hurricane names go back and forth between female names and male names. The list of names is reused every six years. For example, the names for 2018 are the same as those used in 2012.

If a hurricane causes great damage, its name is never used again. Some **retired** hurricane names are Andrew (1992), Mitch (1998), Ivan (2004), Katrina (2005), and Harvey (2017).

Hurricane Andrew caused a great deal of damage, especially in the state of Florida, in 1992.

BIG STORMS

Hurricane Katrina hit the Gulf of Mexico coast in late August 2005. It caused major flooding in New Orleans, Louisiana. With most of the city underwater, tens of thousands of people had to be evacuated. More than 1,800 people died in the storm and its aftermath.

Hurricane Harvey came ashore in Texas on August 25, 2017. When it hit land, its winds topped 100 miles (161 kilometers) per hour. It rained on southeast Texas and Louisiana for the next five days. Some places experienced more than 60 inches (152 centimeters) of rain.

Officials help a family get to dry land after Hurricane Katrina led to major flooding in Louisiana.

THINK ABOUT IT

Hurricane Harvey was just as strong as the Galveston hurricane. What factors might explain why the Galveston hurricane was deadlier than Harvey?

Houston, Texas, located about 50 miles (80 km) from the Gulf of Mexico, was an especially hard-hit area. Thousands of people had to be rescued from their flooded homes. Officials believe Hurricane Harvey caused at least eighty-eight deaths.

One of the deadliest natural disasters in US history was the Galveston hurricane of 1900. This massive storm hit the island city of Galveston, Texas, in September 1900. It claimed more than five thousand lives.

A neighborhood in Houston, Texas, lies mostly underwater following Hurricane Harvey.

HOW DO HURRICANES FORM?

Hurricanes get their power from warmth and moisture that rise from the ocean surface. This power fuels the winds. The winds begin to blow faster and faster. A tropical storm forms when the wind speed gets to 39 miles (63 km) per hour. If the winds blow faster than 74 miles (119 km) per hour, the tropical storm turns into a hurricane.

From above, a hurricane looks like a huge

The red arrows show the warm, moist air that rises from the ocean's surface and powers a hurricane.

COMPARE AND CONTRAST

A wall of clouds, called the eye wall, surrounds the eye. Some of the strongest winds and the heaviest rains are found in the eye wall. What is the difference between the eye and the eye wall?

disk of clouds. Within the storm, high-speed winds move in a spiral around a calm center. The calm center is called an eye. When the eye passes overhead, it can seem as if the storm has ended. However, the opposite side of the storm brings back the strong winds and heavy rain.

top view

eye

eyewall

0 miles   100   200
0 km 100   200   300

The eye is an area of calm in the center of a hurricane. A wall of clouds surrounds the eye.

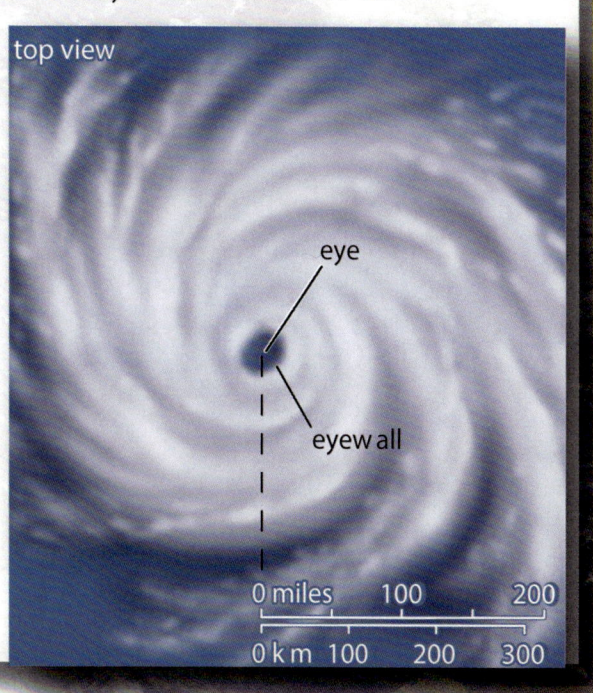

WIND

Wind is the movement of air near Earth's surface. Changes in the temperature of air, land, and water cause wind. When air flows over a warm surface, the air heats up and rises. This action leaves room for cooler air to flow in. The flowing air is wind.

Wind can be a gentle breeze or a strong gale. The most powerful winds happen during storms like tornadoes and hurricanes.

Hurricanes are ranked by the strength of their winds. These ranks are called categories. There are five hurricane categories. They are numbered

Winds from Hurricane Maria battered Puerto Rico in 2017. The National Hurricane Center reported that the winds were close to a category five strength.

COMPARE AND CONTRAST

A tornado is a storm with strong rotating winds. The winds form a funnel that stretches from a cloud toward the ground. The winds of a tornado can reach speeds of up to 300 miles (500 km) per hour. How are tornadoes and hurricanes different? How are they alike?

one through five. A category five is the strongest. It has winds that blow faster than 157 miles (253 km) per hour.

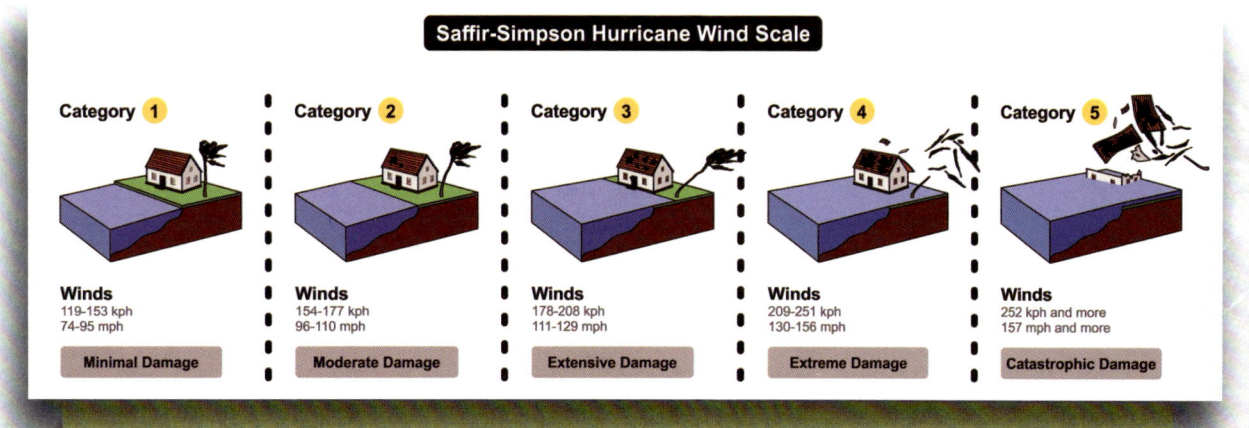

Saffir-Simpson Hurricane Wind Scale

Category 1	Category 2	Category 3	Category 4	Category 5
Winds 119-153 kph / 74-95 mph	Winds 154-177 kph / 96-110 mph	Winds 178-208 kph / 111-129 mph	Winds 209-251 kph / 130-156 mph	Winds 252 kph and more / 157 mph and more
Minimal Damage	Moderate Damage	Extensive Damage	Extreme Damage	Catastrophic Damage

Hurricanes are sorted by their wind speeds. Category five storms have the strongest winds and do the most damage.

WATER

The winds of a hurricane cause a lot of damage, but water is even more destructive. When water covers land that is usually dry, a flood takes place. Flooding is most often caused by heavy rains over long periods of time. Floods are not always bad. They may leave a layer of rich, moist soil when the muddy floodwaters go down. This soil is good for growing crops and other plants. However, when a flood takes place in an area where people live it can cause a lot of damage.

Floodwaters caused by Hurricane Katrina completely covered some New Orleans homes.

THINK ABOUT IT

Think about where hurricanes form. What areas are most in danger of being hit by a hurricane?

Hurricanes can also cause ocean waves to rise to 20 feet (6 meters) above their normal height. The wind pushes the large waves onto shore. The abnormal rise in sea level is called a storm surge. It can cause flooding in cities along the coasts.

Flash floods are the deadliest kind of flood. They are caused by sudden heavy rainfalls. Flash floods happen very quickly and can catch people before they have a chance to escape.

Three years after Hurricane Katrina, Hurricane Gustav caused flooding in parts of New Orleans again.

OCEAN CURRENTS

OCEAN CURRENT

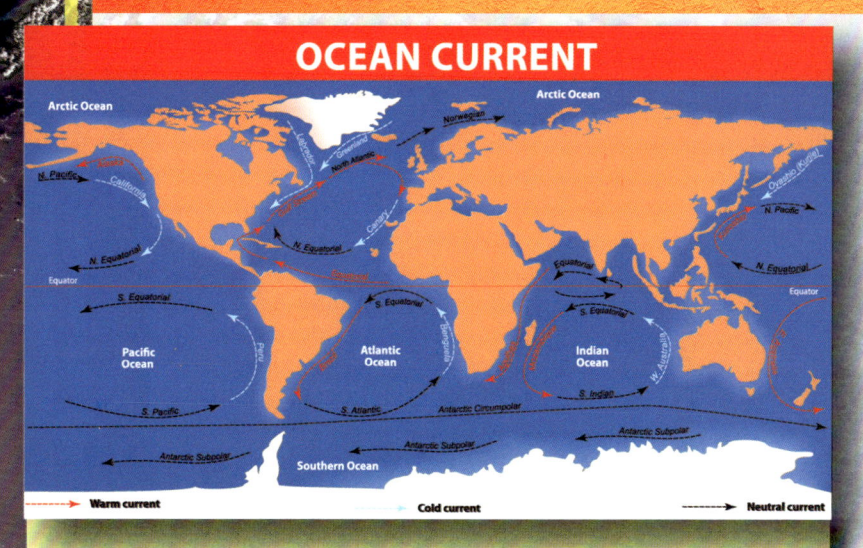

The arrows show major ocean currents and which way they flow.

Winds and other forces cause ocean water to be in constant motion. Large amounts of ocean water move around Earth in patterns called currents. Ocean currents may be warm or cold. The Gulf Stream is a warm current that runs north along the eastern coast of the United States.

A weather condition called El Niño affects the winds and currents in the Pacific Ocean. This weather event occurs every few years in December. In normal years, the winds blow from

COMPARE AND CONTRAST

El Niño and La Niña both result in strong ocean currents. How are El Niño and La Niña different? How are they similar?

east to west this time of year. This wind pattern pushes the warmest water to the west. During El Niño years, the winds slow down. Sometimes, they can even change direction. This shift forces unusually warm water toward the east.

A weather condition called La Niña often follows El Niño. During La Niña, strong winds bring colder than normal waters to the eastern Pacific. Changes in wind patterns and water temperature lead to changes in hurricane activity. Strong winds can tear hurricanes apart and cooler waters stop most hurricanes from forming.

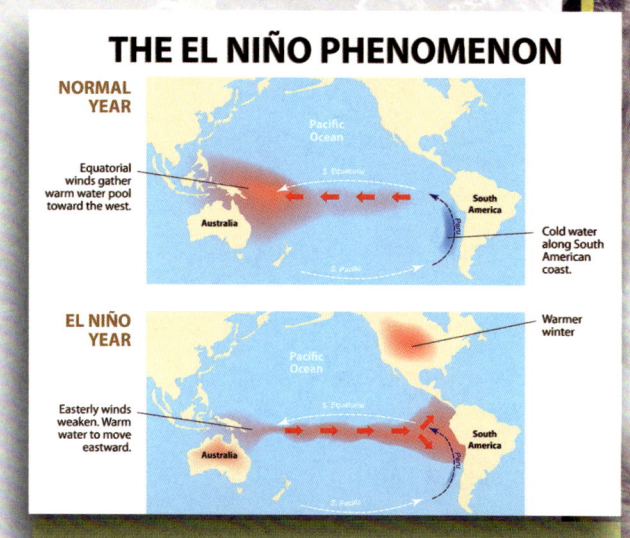

THE EL NIÑO PHENOMENON

Weaker winds during El Niño years allow warm water to move toward South America instead of away from it.

CLIMATE

WORLD CLIMATE REGIONS

Tropical climates: Rainforest, Grasslands, Semiarid, Desert
Subtropical climates: Humid Subtropical, Desert
Temperate climates: Temperate, Subarctic, Grasslands, Desert
Polar climates: Polar Tundra, Polar Ice Cap
Highland climates: Highland (varies with altitude)

© Encyclopædia Britannica, Inc.

Winds and ocean currents are some of the factors that affect climate. Climate is the weather in a certain place over a long period of time. An area's climate determines what kinds of plants can grow and what kinds of animals can live

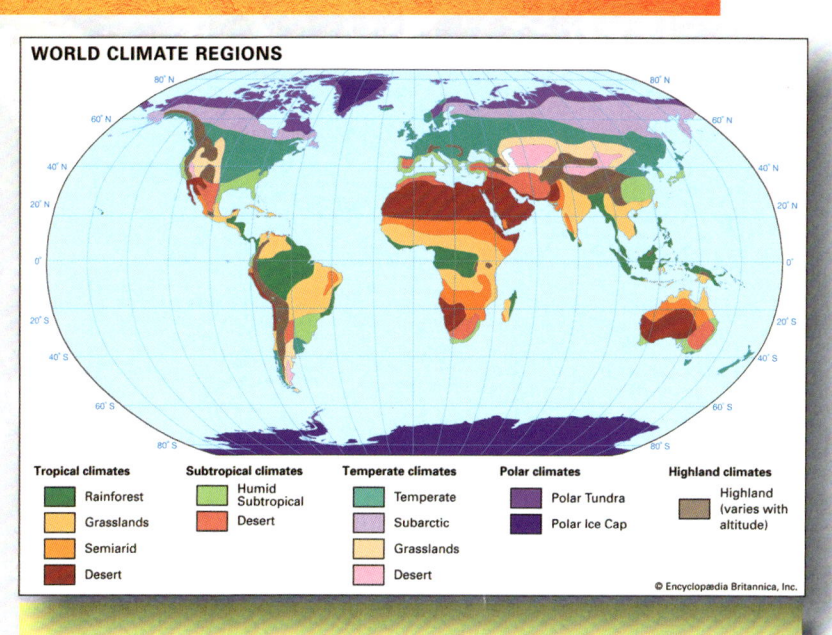

Different areas of the world have different climates. Some climates are hot. Some are cold. Some are dry and others are wet.

COMPARE AND CONTRAST

Sometimes people confuse climate with weather. What is the difference between the two?

there. Some of the other factors that affect climate are the sun, land types, clouds, and human activities.

Ocean currents affect climate because warm currents tend to bring warm weather and rain to nearby land. Cold currents tend to cause a dry climate. In general, land near an ocean has a milder climate than areas that lie inland. The ocean warms the land in winter and cools it in summer. This happens because water heats and cools more slowly than land does.

Land close to water is often warmer in the winter and cooler in the summer than land farther away.

CLIMATE CHANGE

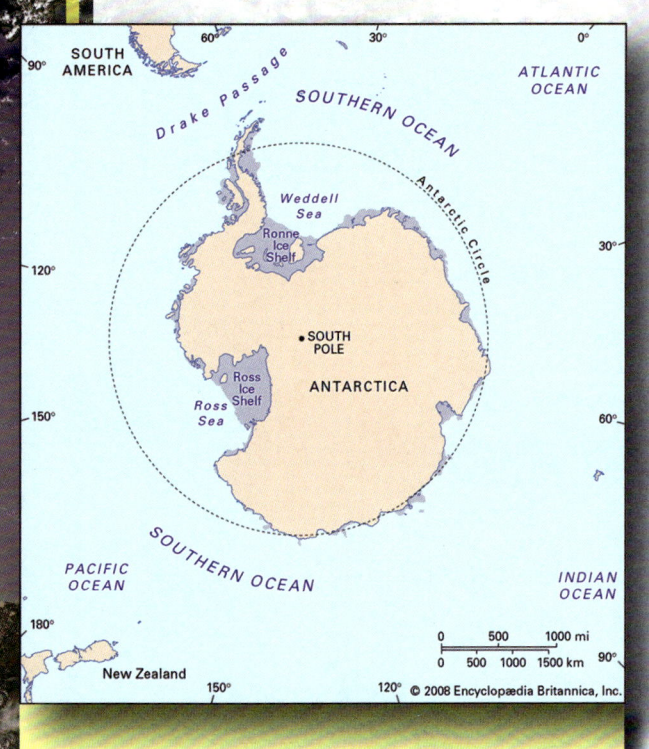

The Southern Ocean surrounds the continent of Antarctica. The South Pole is in Antarctica.

The Southern Ocean (also called the Antarctic Ocean) contains the largest current in the world—the Antarctic Circumpolar Current (ACC). The ACC circles from west to east around Antarctica. It connects the waters from the Indian, Atlantic, and Pacific Oceans and forms a network of ocean currents. These currents spread heat around Earth and thus influence the climate.

The average surface temperature on Earth is slowly

THINK ABOUT IT

Think about where hurricanes get the energy they need to form. If the temperature of Earth continues to rise, what do you think the effect will be on hurricanes?

increasing. This trend is known as global warming. Global warming is the driving effect behind climate change. Climate change is a shift in long-term weather patterns. Some scientists think that the Southern Ocean circulation will slow down as Earth warms. That could speed up the rate of climate change.

To understand global warming, it helps to understand the greenhouse effect. A greenhouse is a glass

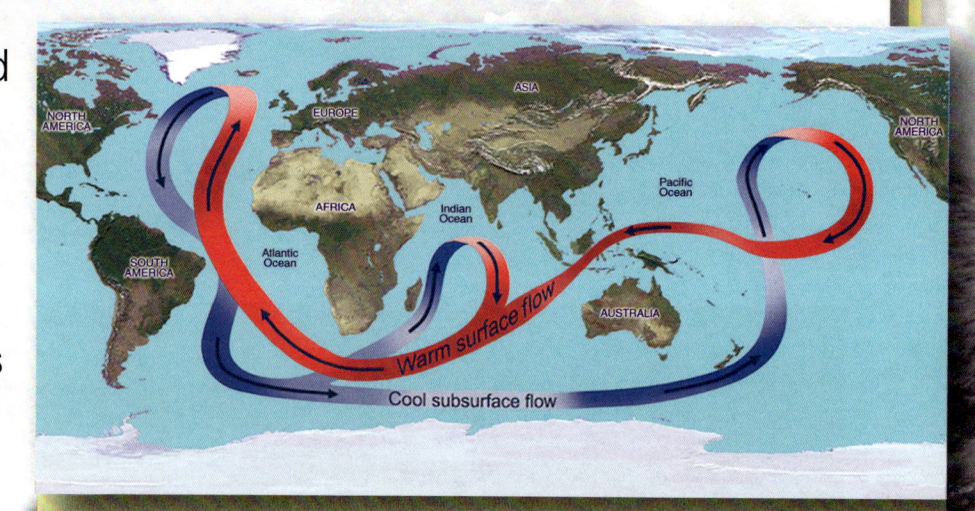

Ocean currents move water all around the world. Water on the surface is warmer than the water below.

house that gardeners use to grow plants in controlled conditions. The glass lets light in and keeps heat from escaping. This trapped heat keeps the plants warm even when it is cold outside.

Likewise, Earth's atmosphere traps energy from the sun and warms the Earth's surface. Land, oceans, and plants absorb, or soak up, energy from sunlight. They release some of this energy as heat. **Greenhouse gases** absorb the

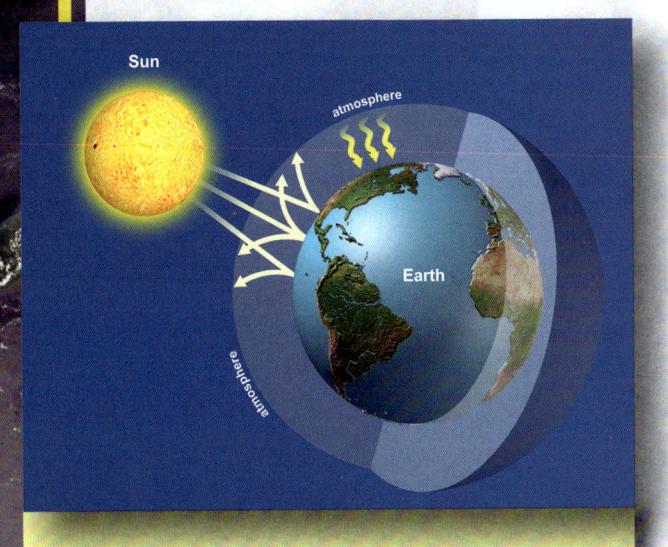

Greenhouse gases in the atmosphere keep heat from the sun from escaping back into space.

VOCABULARY

Greenhouse gases are heat-trapping gases that absorb and trap heat in Earth's atmosphere. The most common greenhouse gases are water vapor, carbon dioxide, and methane.

heat and then send it back toward Earth. Without these gases too much heat would go back into space, and living things could not survive. However, as more greenhouse gases get into the atmosphere, they also trap more heat on Earth. This leads to global warming.

Burning fossil fuels releases greenhouse gases. Power plants burn fossil fuels to make electricity.

For much of Earth's history, greenhouse gases were not a problem. This situation changed as people came to depend on fuels such as oil, gas, and coal. People burn these fuels to power factories, run cars, produce electricity, and heat houses. As the fuels burn, they release carbon dioxide into the atmosphere.

FUTURE HURRICANE SEASONS

Scientists cannot tell how warm Earth may get over time. Some think the temperature may rise between 3.2 and 7.2 degrees Fahrenheit (1.8 and 4 degrees Celsius) by the year 2100. The warmer temperatures could cause climates to change, which could harm living things. For example, polar ice caps might melt. This melting would cause sea levels to rise. Plants, animals, and buildings along the coasts would be in danger.

Polar bears hunt for food on ice. Global warming can cause the ice to melt, so the polar bears may not be able to find enough food to survive.

THINK ABOUT IT

When bad weather threatens, officials work to make sure that people in harm's way are alerted. What can people do to stay out of danger in the future?

It is not yet clear how climate change may affect the number or strength of hurricanes. However, warmer ocean waters would likely lead to stronger storms.

Over time, scientists have gained a better understanding of how hurricanes form and move. This understanding allows them to better forecast where a hurricane will go. People who live in areas most likely to be hit by hurricanes have also become more aware. Most of them have learned to leave the area when experts tell them to do so.

Scientists study climate and weather to understand how hurricanes form and move across oceans.

GLOSSARY

ATMOSPHERE The layer of gases that surrounds Earth.

CATEGORY A grouping of things with common features.

CLIMATE The average weather conditions of a particular place over a period of years.

COMMUNICATE To make something known to another person.

DESTRUCTIVE Causing or tending to cause ruin.

ENERGY Power or the ability to be active.

EYE WALL The area just outside the eye that has the highest winds and heaviest rain.

FORECAST To tell what will happen next based on a study of past data.

FUTURE Coming after the present time period.

GALE A strong movement of air.

MOISTURE A small amount of liquid that causes wetness.

PREDICT To say what will happen next based on what has happened in the past.

ROTATING Turning or causing to turn around a center.

SATELLITE A man-made object or vehicle orbiting Earth, the moon, or another heavenly body.

STORM TRACK The path a hurricane travels along.

TEMPERATURE A measure of hotness or coldness.

WEATHER The daily state of the atmosphere, or air, in any given place.

WEATHER PATTERN Weather that happens over and over in a predictable way.

FOR MORE INFORMATION

Books

Boothroyd, Jennifer. *What Is Severe Weather?* Minneapolis, MN: Lerner Publications Company, 2015.

Jensen, Belinda. *Spinning Wind and Water: Hurricanes.* Minneapolis, MN: Millbrook Press, 2016.

Johnson, Robin. *What Is a Hurricane? (Severe Weather Close-Up).* New York, NY: Crabtree Publishing Company, 2016.

Osborne, Mary Pope. *Hurricane Heroes in Texas* (Magic Tree House). New York, NY: Random House Books for Young Readers, 2018.

Plattner, Josh. *Hurricane or Waterspout?* (This or That? Weather). Minneapolis, MN: Super Sandcastle, 2016.

Spilsbury, Louise, and Richard Spilsbury. *Top 10 Worst Hurricanes* (Nature's Ultimate Disasters). New York, NY: PowerKids Press, 2017.

Websites

National Geographic Kids
https://kids.nationalgeographic.com

National Oceanic and Atmospheric Administration (NOAA)
Hurricanes
http://www.noaa.gov/resource-collections/hurricanes

National Weather Service
Hurricane Safety Tips and Resources
https://www.weather.gov/safety/hurricane

Ready Check
https://www.ready.gov/kids/know-the-facts/hurricanes

SciJinks
https://scijinks.gov

INDEX

Andrew, 11
Antarctic Circumpolar Current (ACC), 24
Atlantic Ocean, 6, 10, 11, 24

China, 5
climate, what it is, 22–23

El Niño, 20–21
evacuate, 8, 9, 12
eye, 15
eye wall, 15

flooding, 5, 12, 13, 18, 19

Galveston, Texas, 13
global warming, 25–27
greenhouse gases, 26–27
Gulf of Mexico, 12, 13

Gulf Stream, 20

Harvey, 11, 12, 13
Houston, Texas, 13
Huracan, 4
hurricanes
    categorizing, 16–17
    characteristics of, 4–5
    how they form, 14–15, 21
    major storms, 12–13
    naming, 10–11
    seasons, 5–6, 28–29
    tracking, 8–9, 28–29

Indian Ocean, 5, 7, 24
Ivan, 11

Japan, 5

Katrina, 11, 12

La Niña, 21

Mitch, 11

New Orleans, Louisiana, 12

ocean currents, 20–21, 22, 23, 24

Pacific Ocean, 7, 10, 20, 21, 24
Philippines, 5

Southern Ocean, 24, 25
storm surge, 19

tornadoes, 17
tropical cyclones, 5, 7
typhoons, 5, 7

winds, and hurricane strength, 16–17